ON BEEKEEPING

THE ANCIENT ROMAN MANUAL
BOOK IX OF *DE RE RUSTICA*

By

Lucius Junius Moderatus Columella

Translated By

Dr. Harrison Boyd Ash

Edited By

Richard Moyer

PREFACE

I now come to the care of wild cattle and the rearing of bees, which I can justly place among creatures which are fed on the farm, since ancient custom placed parks for young hares, wild goats and wild boars near the farm, generally within the view of the owner's dwelling-place, so that the sight of their being hunted within an enclosure might delight the eyes of the proprietor and that when the custom of giving feasts called for game, it might be produced as it were out of store. Also within our own memory, accommodation for bees was provided either in holes cut in the actual walls of the farm building or in sheltered galleries and orchards. So, since we have assigned a reason for the title, which we have prefixed, to this discourse, let us now proceed to deal, one by one, with the topics, which we have proposed.

I.

Wild creatures, such as roebucks, chamois and wild also various kinds of antelopes, deer and wild boars creatures sometimes serve to enhance the splendor and pleasure of their owners, and sometimes to bring profit and revenue. Those who keep game shut up for their own pleasure are content to construct a park, on any suitable site in the neighborhood of the farm buildings, and always give them food and water by hand. Those on the other hand who look for profit and revenue, when there is a wood near the farm (for it is important that it should not be far out of sight of the owner), reserve it without hesitation for the above-mentioned animals, and if there is no natural supply of water, either running-water is introduced or else ponds are dug and lined with mortar to receive and hold the rain-water.

The extent of wood involved is in proportion to the size of each man's property and, if the cheapness of stone and labor make it advisable, certainly a wall built with unhewn stone and lime is put round it; otherwise it is made with unburnt brick and clay. When neither of these methods serves the purpose of the master of the house, reason requires that they should be shut up with a post fence; for this is the name given to a certain kind of lattice made of oak or cork-wood, since olive-wood is only rarely obtainable; in a word, according to local conditions, any kind of wood is chosen for this purpose which resists injury from rain better than any other. Whether it be the round trunk of a tree or cleft into stakes, as its thickness demands, it has several holes bored through its side and is erected firmly in the ground at fixed intervals all round the park; then bars are put across through the holes in the sides of the posts to prevent the passage of the wild beasts. It is enough to fix the posts at intervals of eight feet and to fasten them to the crossbars in such a way that the width of space, which occurs where holes are left, may not offer the animals a means of escape. In this manner you can even enclose very wide regions and tracts of mountains, as the vast extent of ground permits in the provinces of Gaul and in certain others; for there is both a great abundance of timber for making posts and everything else which is needed for the purpose is in abundant supply. The soil abounds in frequent springs, which is particularly wholesome for the above-named kinds of animals; then too it furnishes wild creatures with fodder most liberally even of its own accord. Woodlands are chiefly chosen which abound in the fruits of the ground and also in trees; for as these creatures have need of grass, so too they require the fruits of oak-trees, and those woods are most highly commended which are most productive of the acorn of the ordinary oak and of the evergreen oak and likewise of the Turkey-oak, also of the fruit of the strawberry-tree and the other wild fruits which we described in great detail when we were discussing farm-yard pigs.

Nevertheless the careful head of a household ought not to be content with the foods which the earth produces by its own nature, but, at the seasons of the year when the woods do not provide food, he ought to come to the help of the animals which he has confined with the fruits of the harvest which he has stored up, and feed them on barley or wheat-meal or beans, and especially, too, on grape-husks; in a word, he should give them whatever costs the least. Also in order that the wild creatures may understand that provision is being made for them, it will be a good plan to send among them one or two animals which have been tamed at home, and which, roaming through the whole park, may direct the hesitating creatures to the fare offered to them. It is advisable that this should be done not only during the scarce season of winter but also when those which were with young have brought them forth, so that they may rear them better. And so the park-keeper will have frequently to be on the watch and see if they have borne their young, in order that cereals given them by hand may sustain their strength. But neither the antelope nor the wild boar nor any other wild creature should be allowed to live to a greater age than four years. For up to that time they advance in growth, after it they grow old and lean; and so they should be turned into cash while a vigorous time of life preserves their bodily comeliness. The deer, however, may be kept for many years, for it long remains young in your possession, because it has been allotted a life of longer duration. But as regards animals of lesser growth, such as the hare, our advice is that, in those parks surrounded by a wall, the seeds of mixed cereals and of the pot-herbs, wild endive and lettuce, should be thrown upon small beds of earth made at different intervals apart. Also the Carthaginian and our own native chick-pea, and barley too and chickling should be produced out of store and put before them after having been soaked in rain-water; for dry food is not much sought after by hares. Moreover, it is easily understood even without my mentioning it, concerning these animals and others like them, how inexpedient it is to introduce them into a park which is surrounded by railings, since owing to the small size of their bodies they can easily creep under the bars and, having obtained free exit, manage to escape.

II.

I come now to the management of beehives about which no instructions can be given with greater care than in the words of Hyginus, more ornately than by Vergil, or more elegantly than by Celsus. Hyginus has industriously collected the opinions of ancient authors dispersed in their different writings; Vergil has embellished the subject with the flowers of poetry; and Celsus has applied the method of both the above-mentioned authors. Therefore, we ought never to have even attempted to discourse on this subject, did not the fulfillment of the promise which we made call for the treatment of this part of our subject also, lest the body of the work begun, looked at as a whole, should appear mutilated and imperfect, as if a limb had been cut off. The tradition of the fabulous origin of the bees which Hyginus has not passed over I would rather attribute to poetic license than submit to the test of our belief; nor indeed is it a fit question for a husbandman to ask whether there ever existed a woman of surpassing beauty called Melissa, whom Jupiter changed into a bee, or whether (as Euhemerus the poet says) the bees were bred from hornets and the sun, and that the nymphs, the daughters of Phryxon, reared them, and that soon after they became the nurses of Jupiter in the Dictaean Cave and that, by the gift of the god, they had allotted to them the food with which they themselves had reared their little foster-child. Upon this story, though not unworthy of a poet, Vergil touched briefly and lightly in a single line when he said:

"Neath Dicte's cave they fed the king of heaven."

But it does not even concern husbandmen when and in what country bees first came into existence, whether in Thessaly under Aristaeus, or in the island of Cea, as Euhemerus writes, or on Mount Hymettus "in the time of Erechtheus," as Euthronius says, or in Crete in the time of Saturn, as Nicander says. All this no more concerns farmers than the question whether the swarms of bees produce their offspring, as we see the other animals do, by copulation, or whether they pick up the heir of their race from the flowers, as our own poet Maro affirms, and whether they vomit the liquid honey from their mouths or yield it from some other part. The inquiry into these and similar questions concerns those who search into the hidden secrets of nature rather than husbandmen. They are subjects more agreeable to the students of literature, who can read at their leisure, than to farmers who are busy folk, seeing that they are of no assistance to them in their work or in the increase of their substance.

III.

Therefore let us return to topics that are more suitable to those who have charge of beehives, how many kinds of bees there are and which of them is the best. Aristotle, the founder of the Peripatetic School, in the books which he wrote about animals, shows that there are several kinds of swarms of bees, some of them having bees huge and globular in shape and at the same time black and hairy; others smaller but equally round and of a dusky color and with bristling hairs; others still smaller but not so round, but nevertheless fat and broad and of rather a better color; some very small and slender with bellies which end in a point, striped of a golden color and quite smooth. Vergil, following Aristotle as his authority, approves most of bees which are very small, oblong, smooth and shining, "burning with gold, their bodies stained with spots of equal size," calm, too, in disposition; for the larger and rounder a bee is, the worse it is, and if it is unusually fierce, it is by far the worst kind of all. However, the irascibility of the better kind of bees is easily soothed by the frequent intervention of those who look after them; for when they are often handled, they quickly become tame. If they are carefully looked after, they live for ten years; but no swarm can exceed this age, even if young stock is substituted yearly in place of those that have died; for usually in the tenth year all the population of the whole hive is destroyed and exterminated. In order, therefore, that this may not be the fate of the whole apiary, fresh stock must be continually propagated and care must be taken in the spring, when the fresh swarms issue forth, that they are intercepted and the number of dwelling places increased; for bees are often overtaken by diseases. The methods by which these ought to be cured will be dealt with in their proper places.

IV.

Meanwhile, when you have chosen your bees in accordance with the points which we have just mentioned, feeding-grounds ought to be assigned to the bees of which you approve. These should be as retired as possible and, as our Maro directs, void of cattle and with a sunny aspect as little as possible exposed to storms, where winds may not approach; for winds prevent the bees from bearing home their food; nor sheep. Nor frisky kids must trample down the flowers. Nor heifers wandering o'er the plain shake off the dews or crush the rising blades of grass.

The region should also be rich in small clumps, especially thyme and marjoram and also in Greek savory and our own Italian savory, which the country-folk call satureia. Next let there be plenty of shrubs of larger growth, such as rosemary and both kinds of trefoil (for there is one variety which is sown and another which grows of its own accord), also the evergreen pine and the lesser holm-oak (for the taller variety is universally condemned). Ivy, too, is admitted not for its other good qualities but because it provides a large quantity of honey. Of trees the following are very highly commended, the red and white jujube-trees, likewise tamarisks, also almond-trees and peach-trees and pear-trees, in a word, so as not to waste time in naming each kind, the majority of the fruit-bearing trees. Of woodland trees the most suitable are the acorn-bearing oaks, also terebinths and mastic trees, which closely resemble them, and lime-trees. Of all the trees of this class yews only are excluded as being hurtful. Moreover a thousand seeds, which flourish in uncultivated turf or are turned up in the furrow, produce flowers which are much loved by bees, for example shrubs of starwort in virgin soil, stalks of bear's foot, stems of asphodel and the sword-like leaf of the narcissus. White lilies sown between the furrows in the garden make a brilliant show and the gillyflowers have no less pure a color; then there are red and yellow roses and purple violets and sky-blue larkspur; also the saffron-bulbs are planted to give color and scent to the honey. Moreover, countless herbs of a baser kind spring up on cultivated land and pasture which supply an abundance of wax for the honey-combs, such as the common charlock and the horse-radish, which is no more precious, the mustard-herb, and flowers of wild endive and black poppy, also the field parsnip, and the carrot. But of all the plants that I have suggested and of those that I have not mentioned so as to save time (for their number could not be computed), thyme yields honey with the best flavor; the next best are Greek savory, wild thyme and marjoram. In the third class, but still of high quality, are rosemary and our Italian savory, which I have called satureia. Next the flowers of the tamarisk and the jujube-tree and the other kinds of fodder that I suggested have only a mediocre flavor. The honey that is considered of the poorest quality is the woodland honey that comes from dirty feeding-grounds and is produced from broom-trees and strawberry trees, and the farmhouse honey that comes from vegetables. Now that I have described the situation of the feeding grounds and also the various kinds of food, I will next speak of the arrangement for receiving and housing the swarm.

V.

A position must be chosen for the bees facing the sun at midday in winter, far from the noise and the assemblage of men and beasts and neither hot nor cold, for bees are troubled by both these conditions. It should be situated in the bottom of a valley, that the empty bees, when they go forth to feed, may be able more easily to fly up to the higher ground, and also, when they have collected what they require, they may fly with their burden on a down-hill course without any difficulty.

If the situation of the farm permits, we ought not to hesitate to join the apiary to a building and surround it with a wall, but it must be on the side of the house that is free from the foul odors that come from the latrines, the dunghill and the bathroom. If, however, this position has drawbacks, but yet the worst disadvantages are not all present, even under these conditions it will be more expedient for the apiary to be under the master's eye. If, however, everything is unfavorable, at all events a valley should be pitched upon close at hand, so that the owner may be able to go down rather often and visit it without grave inconvenience; for in bee-keeping perfect honesty is necessary, and since this is very rare, it is better secured by the intervention of the master. Not only is an overseer who is fraudulent abhorrent to the business but also one whose laziness causes filthy conditions; for beekeeping revolts alike against a lack of cleanliness and against fraudulent management.

Wherever the hives are placed, they should not be

enclosed within very high walls. If, through fear of robbers, a rather lofty wall is thought desirable, passages through it should be made for the bees in the form of a row of little windows three feet above the ground, and there should be an adjoining cottage in which the keepers may live and the apparatus may be stored. The storehouse should be chiefly occupied by hives ready for the use of new swarms and also by health-giving herbs and any other remedies that may be applied to bees when they are sick.

> *"And let a palm or vast wild-olive tree*
>
> *O'ershade the porch, that when new kings lead forth*
>
> *The infant swarms and the young bees make sport*
>
> *In their own spring, from honeycombs set free;*
>
> *Then let the neighboring bank invite retreat*
>
> *From mid-day heat, and let the sheltering tree*
>
> *Hold them in leafy hospitality."*

Next let ever-flowing water, if it is available, be introduced or drawn by hand and provided, without which neither combs nor honey nor even young bees can be formed. Whether, therefore, as I have said, it is running water that has been conveyed in channels or well water, it should contain heaps of sticks and stones for the use of the bees,

That upon frequent bridges they may rest and spread their wings to catch the summer sun, if swift east winds have caught them loitering and rained on them or plunged them in the deep.

Next, round the whole apiary, little trees of small growth ought to be planted and in particular shrub trefoils on account of their health-giving properties (for they are a remedy for bees when they are listless); also wild cinnamon and pines and rosemary, and clumps of marjoram and thyme and violets and whatever else the nature of the ground allows to be profitably planted. Not only growing things but also anything whatsoever which has a disagreeable and noisome odor should be kept away from the apiary, such as the smell of a crab when it is burnt on the fire or the odor of mud taken from a marsh. Likewise let hollow rocks and shrill noises produced by valleys, which the Greeks call echoes, be avoided.

VI.

When, therefore, the sites have been arranged, the beehives must be constructed in accordance with local conditions. If the place is rich in cork-trees, we shall certainly make the most serviceable hives from their bark, because they are neither cold in winter nor hot in summer; or if it grows plenty of fennel-stalks, with these too, since they resemble the nature of bark, receptacles can be quite as conveniently made by weaving them together. If neither of these materials is at hand, the hives can be made by plaiting withies together; or, if these are not available either, they will have to be made with wood of a tree either hollow or cut up into boards. Those made of earthenware have the worst qualities of all, since they are burnt by the heat of summer and frozen by the cold of winter. Two kinds of hives remain to be described, those which are either made of dung or built of bricks. Celsus was right in condemning the former because it is very liable to catch fire; the latter he approved, although he made no secret of its chief disadvantage, namely, that if occasion should arise, it cannot be moved to another site. I do not agree with him who thinks that hives of this kind ought to be used in spite of this drawback, for it is not only against the interests of the owner that they should be immovable when he wants to sell them or furnish another site with hives (for these considerations concern the convenience of the owner alone), but the question arises as to what ought to be done for the sake of the bees themselves, when it is advisable that they should be sent to another district because they are suffering from disease or from the barrenness and poverty of the locality and yet cannot be moved for the reason mentioned above — a state of affairs which ought above all things to be avoided. So, though holding in respect the authority of a learned man, yet, without seeking to set myself up against him, I have not omitted to express my own opinion. For Celsus' chief anxiety, lest the bees' quarters should be exposed to fire or thieves, can be avoided by building a brick wall round the hives to prevent the plundering of robbers and to give protection against the violence of fire, and, when the hives have to be moved it will be possible to take apart the framework of the structure and move the hives elsewhere.

VII.

But since most people regard all this as involving too much trouble, whatever kind of receptacles beehives take their fancy will have to be arranged thus. A bank made of stones built three feet high is stretched across the apiary and carefully smoothed over with plaster, so that no chance of climbing it may be offered to lizards and snakes or other harmful creatures; then on the top of it are placed either bee-houses made with bricks, which Celsus prefers, or, as we prefer, hives walled round except at the back; or else — and this is the practice of almost all those who are careful in these matters — receptacles arranged in a row are fastened together either with small bricks or with unhewn stones in such a way that each is contained within two narrow walls and the two sides, at the back and at the front, are left free; for the sides on which they issue forth have sometimes to be opened and this is especially necessary at the back because the swarms have to be attended to from time to time. If there are no partitions between the hives, they will, nevertheless, have to be so placed as to be at a little distance from one another, so that, when they are being inspected, one which is handled in the course of being attended to may not shake another which is closely joined to it, and alarm the neighboring bees, which are afraid of every movement as threatening ruin to their structures which are frail, being of wax. It is quite enough to have three rows of hives one above the other, since even so the man who looks after them cannot very conveniently inspect the top row. The fronts of the hives, which afford entries for the bees, should slope down more than their backs, so that the rain may not flow in, and that, if by chance it does find its way in, it may not remain there but flow out through the entrance. Also, on account of the rain, the hives should be protected above with colonnades, or, failing these, they should be overshadowed by green foliage daubed over with clay, forming a covering that keeps off both the cold and rain and also the heat. However the heat of summer is not so harmful to this kind of creature as the cold of winter, and so there should always be a building behind the apiary to intercept the violence of the north wind and provide warmth for the hives. Likewise the bees' dwelling-places, although they are protected by buildings, ought to be so arranged as to face the south-east, in order that the bees may enjoy the sun when they go out in the morning and may be more wide-awake; for cold begets sloth. For the same reason, too, the holes through which they go in and out ought to be very narrow, so as to admit as little cold as possible; indeed it is enough that they should be so bored that they cannot admit more than one bee at a time. Thus neither the poisonous gecko nor the foul race of beetles and butterflies and the cockroaches that shun the daylight, as Maro says, will lay waste to the honeycombs by having too wide an entrance to pass through. It is most useful to have made in proportion to the number of bees in the hive, two or three entrances in its outer covering at a distance from one another to defeat the craftiness of the lizard, which standing like a door-keeper at the entry, with open mouth, brings destruction upon the bees as they come forth, and fewer of them perish when they are at liberty to avoid the pest which lies in wait for them by flying out by another passage.

VIII.

We have now said enough about the choice of feeding-grounds, dwelling-places and their sites, and these having been provided, the next things that we require are swarms or bees. These come to us either by purchase or without being paid for. Those which we are going to buy we shall test with particular care by means of the points already mentioned, and we must consider how numerous they are before we purchase them, by opening the hives; or if there are no facilities for inspecting them, we shall at any rate take note of what we are allowed to see, namely, whether a goodly number of bees are standing in the entrance-porch and whether a loud noise is to be heard of bees buzzing inside. Also if it so happens that they are all silent and at peace within their dwelling-place, we shall be able to estimate their great or small number from the sudden noise on the part of the bees as a result of our applying our lips to the hole by which they enter and blowing into it.

But we must be particularly careful that the swarms are brought from the neighborhood rather than from distant regions, since they are usually irritated by a change of climate. But if this is impossible and we are obliged to convey them over long distances, we shall be careful that they are not disturbed by the roughness of the road, and they will be best carried on the shoulders and at night; for they must be given rest in the day-time, and liquids which they like must be poured into the hives, so that they may be fed while remaining shut up. Then when they have arrived at their destination, if daylight has come on, the hive must be neither opened nor placed in position until evening conies, so that the bees may go forth quietly in the morning after a whole night's rest, and we shall need to watch carefully for about three days to see whether they all sally forth in a body; for when they do this, they are meditating escape. We will presently prescribe what remedies we ought to apply to prevent this. Bees which come to us by gift or by capture are accepted less scrupulously, although even in these circumstances I would not care to possess any but the best, since good and bad bees require the same expenditure and the same labor on the part of their keeper; also (and this is especially important) inferior bees should not be mixed with those of high quality, since they bring discredit upon them; for a smaller yield of honey rewards your efforts when the idler swarms take part in the gathering of it. Nevertheless, since sometimes, owing to local conditions, an indifferent set of bees has to be procured (though never on any account should a bad one be acquired), we shall exercise care in seeking out swarms by the following method. Wherever there are suitable woodlands where honey can be gathered, there is nothing that the bees would sooner do than make choice of springs near at hand for their use. It is a good plan, therefore, usually to frequent these springs from the second hour onwards and watch how many bees come to them for water. For if only a few are flying about (unless there are several sources of water which attract them and cause them to be more widely dispersed) we must conclude that there is a scarcity of them, which will make us suspect that the place will not produce much honey. But if they come and go in large numbers, they inspire greater hopes of our catching swarms of them; and the following is the method of finding them. First we must try to discover how far away they are, and for this purpose liquid red-ochre must be prepared; then, after touching the backs of the bees with stalks smeared with this liquid as they are drinking at the spring, waiting in the same place you will be able more easily to recognize the bees when they return. If they are not slow in returning, you will know that they dwell in the neighborhood; but if they are late in doing so, you will calculate the distance by the period of their delay. If you notice them returning quickly, you will have no difficulty in following the course of their flight and will be led to where the swarm has its home. As regards those who apparently go farther away, a more ingenious plan will be adopted, as follows. The joint of a reed with the knots at either end is cut and a hole bored in the side of the rod thus formed, through which you should drop a little honey or boiled-down must. The rod is then placed near a spring. Then when a number of bees, attracted by the smell of the sweet liquid, have crept into it, the rod is taken away and the thumb placed on the hole and one bee only released at a time, which, when it has escaped, shows the line of its flight to the observer, and he, as long as he can keep up, follows it as it flies away. Then, when he can no longer see the bee, he lets out another, and if it seeks the same quarter of the heavens he persists in following his former tracks. Otherwise he opens the hole and allows them to emerge one after another, and marks the direction in which most of them fly home, and pursues them until he is led to the lurking-place of the swarm. If it is hidden in a cave, the swarm will be driven out with smoke, and when it has sallied forth, it is checked by the noise of brass being beaten; for, terrified by the sound, it will immediately settle on a shrub or on a higher kind of foliage, that of a tree, and is enclosed in a vessel prepared for the purpose by the man who has tracked down the bees. But if the swarm has its home in a hollow tree and either the branch which the bees occupy stands out from the tree or they are inside the trunk of the tree itself, then, if the small size of the branch or trunk allows, first the upper part, which is empty of bees, is cut through with a saw which should be very sharp so that the operation may be more quickly carried out, and then the lower part so far as it seems to be inhabited. Then, when it has been cut off at both ends, it is covered with a clean garment (for this too is very important), and if there are any gaping holes," they are daubed over, and it is carried to the place where the bees are kept, and, small holes being left in it, as I have said, it is put in position like the rest of the hives. The searcher for swarms should choose the morning for his search, so that he may have the whole day to spy out the comings and goings of the bees. For often, if he is too late in beginning to observe them, when they have finished their usual tasks, they go home and do not return to the water, even though they are near at hand, with the result that the man who is searching for them does not know how far away the swarm is from the fountain.

Gather bruised balm and wax-flower's lowly greenery, and other similar herbs in which this kind of creatures takes delight, and rub the hives thoroughly with them, so that the scent and juice stick to them; then, after cleaning them, they sprinkle them with a little honey and place them here and there in the woods not far from the springs and, when they are full of swarms, they carry them back home. It is not, however, expedient to do this except in places where there is an abundance of bees, because it often happens that chance passers-by, finding the hives empty, carry them off with them, nor is the possession of one or two full of bees enough to compensate for the loss of several empty hives. But where bees are more plentiful, even if many hives are carried off, the bees that are obtained make up their loss. Such is the method of catching wild swarms of bees.

IX.

Next there is another method of retaining the swarms produced from our own apiaries. The keeper ought always diligently to go round the hives, for there is no time when they do not need his care but they demand still more careful attention when the bees feel the approach of spring and the hives overflow with new offspring, which, unless they are promptly intercepted by the constant watchfulness of the keeper, fly off in different directions. For such is the nature of bees that each brood of ordinary bees is generated together with its king and, when they have acquired enough strength to fly away, they despise the society of their elders and even more the orders which they give; for as the human race, which possesses reason, allows no partnership of the kingly power, much less do the dumb animals who are lacking in understanding. Therefore the new chieftains come forth with their following of young bees, which, remaining in a mass for one or two days at the very entrance of their abode, by their coming out show their desire for a home of their own, and if the man in charge immediately assigns it to them, are as content with it as if it were their native place. If, however, the keeper has been away, they make for some strange region as if they had been driven away unjustly. To prevent this, it is the duty of a good overseer in springtime to keep an eye upon the hives until about the eighth hour of the day (after which the new battalions of bees do not take to impetuous flight), and carefully watch their departures, for some of them, when they have broken out, usually immediately rush away. He will be able to find out beforehand their decision to escape by putting his ear to each of the hives in the evening; for about three days before they intend to break out an uproar and buzzing arises like that of an army setting out on the march. From this, as Vergil very truly says.

> "You can foreknow the purpose of the herd;
>
> The martial roar of the hoarse brass reproves
>
> The loiterers, and a voice is heard whose notes
>
> The broken sound of trumpets imitates."

The bees, therefore, which behave like this ought especially to be kept under observation, so that, whether they sally forth to battle (for they wage a kind of civil war amongst themselves and as it were foreign wars with other swarms) or break out in order to escape, the keeper may be at hand, ready for either event. Fighting either of the bees of one swarm quarrelling amongst themselves or of two swarms at variance with one another is easily quelled; for, as the same poet says,

> "By casting of a little dust the strife is stayed and laid to rest,"

or else by sprinkling over them honey-water or raisin-wine or some similar liquid, that is to say the sweet taste of things familiar to them, abates their wrath. The same expedients too are wonderfully efficacious for reconciling king-bees when they are at enmity; for there are often several leaders of one people, and the common herd is as it were divided into factions by the quarrels of its chiefs. This must be prevented from happening often, since whole nations are destroyed by civil war. And so, if good feeling exists between the princes, peace continues and no blood is shed. If, however, you have often noticed them fighting a pitched battle, you will take care to put to death the leaders of the factions; but when they are actually fighting, the above-mentioned remedies can calm their battles. Next, when a host of bees has settled in a mass on the neighboring branch of a leafy shrub, you should take notice whether the whole swarm hangs down in the form of a single bunch of grapes. This will be a sign either that there is only one king-bee in it or, at any rate, that, if there are several, they are reconciled and on good terms with one another, in which case you will leave them as they are until they fly back to their abode. If, however, the swarm is divided into two or even more clusters, you need have no doubt that there are several chiefs and that they are still in an angry mood, and you will have to search for the leaders in the parts of the clusters where you see the bees most closely massed together. Having, then, smeared your hand with the juice of the herbs already named, that is, balm and wild parsley, lest they fly away at your touch, you will gently insert your fingers and, after separating the bees from one another, you will search until you find the author of the quarrel.

X.

Now the king-bees are slightly larger and more oblong in shape than the other bees, with straighter legs but less ample wings, of a beautiful shining color and smooth, without any hair, and stingless, unless one regards as such the coarser hair-like object growing on their belly, of which, however, they do not make use to inflict a hurt. Some, too, are found of a dusky color and hairy, of whose disposition you will form an unfavorable opinion judging from their bodily appearance.

> *"As two-fold are the features of the kings,*
>
> *So are the bodies of their subjects; one*
>
> *Will gleam with markings rough with gold, and bright*
>
> *With ruddy scales, and of a comely mien."*
>
> -Vergil

That is why this one is especially approved, being superior; for the inferior kind, like dirty spittle, is as foul as the wayfarer who comes from depth of dust and from his parched mouth the dirt spits forth:

And as the same writer says,

> "*With sloth inglorious his wide paunch he drags.*
>
> *Therefore all the leaders of the baser kind*
>
> *Give them to death, and let the better prince*
>
> *Rule in the empty hall."*

Nevertheless he too must be despoiled of his wings, when he oft-times attempts to break out with his swarm and fly away; for, if we strip him of his wings, we shall keep the vagrant chieftain as though in fetters chained, who, deprived of the resource of flight, ventures not to leave the confines of his realm and, for this reason, does not allow even the people under his sway to wander further than he is able.

XI.

But sometimes the king-bee has to be put to death when an old hive falls short of its proper complement of bees, and its want of numbers must be made up from another swarm. Therefore, when in the early spring a young brood is born in the hive, the new king-bee is squeezed to death, so that the multitude of bees may live with their parents without discord. But if the combs have produced no offspring, it will be open to you to bring together the population of two or three hives into one, but only after they have been sprinkled with sweet liquid; then you can shut them up and, after placing food for them, keep them enclosed for about three days leaving only small breathing-holes, until they are accustomed to live together. There are some people who prefer to get rid of a king-bee that is old, but this is harmful; for the crowd of older bees, who form a kind of senate, do not think fit to obey the juniors and, through obstinately despising the orders of those who are stronger than themselves, are visited with punishment and death. The trouble, indeed, which usually befalls a younger swarm, when the king of the old bees whom we have left in power has failed through old age and wild discord arises through lack of control (just as happens in a family when its head dies), can easily be met. For one leader is chosen from those hives which have several chiefs and is transferred to those which have no one to govern them, and set up as ruler. In those quarters that are afflicted by some pestilence the lack of bees can be remedied with less trouble; for when the disaster to the crowded hive is recognized, you must examine any combs that it contains. You must then next cut away, from the wax that holds the seeds, that part in which the offspring of the kingly race comes to life. It is easy to see this, since almost at the very end of the wax there appears as it were the nipple of a breast projecting somewhat and with a wider cavity than the rest of the holes, in which the young bees of the common kind are enclosed. Celsus indeed declares that there are five transverse cavities in the outermost combs that contain the royal progeny. Hyginus, too, following the authority of the Greeks, says that the ruler is not formed, like the rest of the bees, from a small worm, but that, on the circumference of the combs, straight holes are to be found somewhat larger than those which hold the bees of common birth, filled with a kind of dirt of a red color from which the winged king-bee is immediately formed.

XII.

Care must also be taken of the home-bred swarm, if by chance, taking a dislike to their paternal swarm and abode, they break forth at the time already mentioned prevent it and announce their intention of taking a more distant flight. This the swarm intimates when the bees so completely avoid the entrance to the hive that not a single one flies back again into it, but immediately rises high into the sky. The young bees who are escaping should be frightened by the rattling of brass or potsherds, which are usually to be found lying about; and when in their alarm they have returned to the maternal hive and hang in a mass at the entrance to it or betake themselves immediately to the nearest foliage, the keeper should immediately besmear the inside of a new receptacle prepared for the purpose with the herbs mentioned above, and then, after sprinkling it with drops of honey, bring it near and gather the mass of bees together with his hands or with a scoop; and, after taking every proper precaution, he should let the hive, after it has been carefully adjusted and besmeared inside, remain in the same place until evening begins to fall. Then at first twilight he should remove it and replace it in a row with the other hives. But you should also have empty hives placed in the apiary; for there are some swarms which, as soon as they have come forth, immediately seek a home for themselves nearby and occupy one that they find empty. You now have a practically complete account of the measures to be taken for acquiring bees and keeping them in your possession.

XIII.

The next thing is that remedies are needed for those that are suffering from disease or pestilence. The ruinous disease of pestilence is rare in bees, nor can I find anything which ought to be done other than what we have prescribed in the case of the other animals (except that the hives should be moved far away); but the causes of common ailments in bees are more easily diagnosed and remedies found for them. The most serious is their annual distemper at the beginning of spring, when the spurge-bush flowers and the elms put forth their bitter blossoms; for as by fresh apples, so are they allured by these early flowers and eat greedily of them after their winter hunger, such food not being hurtful when not eaten beyond satiety, but when they have gorged themselves abundantly with it, they die from a flux of the belly, unless help is quickly given. Spurge produces looseness of the bowels in the larger animals as well, but elm has this effect particularly on bees. This is the reason why bees rarely continue numerous in the districts of Italy that are planted with trees of this kind. And so at the beginning of spring, if you supply them with medicated food, by means of the same remedies it is possible both to provide against their being troubled by plague of this kind and also to cure them when they are already suffering from it. Now I myself do not venture to insist on the treatment that Hyginus, following ancient authorities, has recorded, since I have not tried it; but it is open to those who wish to do so to test it. For his instructions are: when a plague of this kind has attacked the bees, and the bodies are found for dead in heaps under the honeycombs, lay them aside in a dry place through the winter, and, at about the time of the spring equinox, when the mildness of the day invites us, bring them out into the sunshine, after the third hour, and cover them with fig-wood ashes. If this is done, he declares that within two hours, brought to life by the quickening breath of the heat, they begin to breathe again and crawl into a vessel provided for this purpose, if it is placed in their way. We rather, that they may not perish, are of opinion that the diet, which we will forthwith describe, should be put before the swarms when they are sick. For they ought to be given either seeds of pomegranate, bruised and sprinkled with Aminean wine, or raisins with an equal quantity of Syrian sumac and soaked in rough wine; or, if these are without effect taken separately, all the same ingredients should be pounded in equal quantities into a single mass and boiled in an earthenware vessel with Aminean wine and then allowed to cool right away and placed before the bees in wooden troughs. Some people boil rosemary in honey-water and, when it has cooled, pour it into troughs and give it to the bees to sip. Others put the urine either of oxen or of human beings near the hives, as Hyginus declares. Moreover also, that disease is particularly remarkable which makes them hideous and shrunken and consumes them, when some often carry out from their abodes the bodies of those which have died, while others remain listless within their dwellings in sad silence, as though in time of public mourning. When this happens food is offered them poured into troughs made of reeds, especially boiled honey pounded up with an oak apple or a dried rose. It is also a good plan to burn galbanum, that they may be cured by its odor, and to keep up their strength, when they are exhausted, with raisin-wine and boiled-down must. The root of the starwort, the bushy part of which is yellow and its flower purple, has the best effect of all; it is boiled with old Aminean wine and pressed and then the juice is strained and given as a remedy. Hyginus indeed, in the book which he wrote about bees, says: "Aristomachus is of opinion that help ought to be brought to bees which are sick in the following manner: first, all the diseased combs should be removed and entirely fresh food placed for the bees, and then they should be fumigated." He thinks also that it is beneficial to add a new swarm to the bees that are wasted by old age, although there is a danger that they may be destroyed by sedition, nevertheless they are likely to rejoice because their number is increased. But that they may remain in a state of concord, the kings of those bees that are being transferred from another hive ought to be put out of the way as rulers of an alien people. There is, however, no doubt that the honey-combs of the most populous swarms, which have young bees already matured in them, ought to be transferred and made subject to the less populous swarms that their families may be strengthened by the adoption, as it were, of fresh progeny. But, when this is going to be done, we must remember to put in the care of the old swarm those honey-combs in which the young ones are already opening their cells and putting out their heads and eating away the wax which was laid upon the top as a kind of covering for their holes. For if we transfer the honeycombs when the brood has not come to maturity, the young bees will die when they cease to be kept warm. For they often die of a distemper which the Greeks call phagedina. For since it is the habit of bees to construct beforehand as many cells as they think they can fill, it sometimes happens that, when their waxen structures are finished, the swarm, while it is roaming too far afield in search of honey, is overwhelmed in the woods by sudden showers and whirlwinds and loses most of the ordinary bees. When this has happened, the few that remain are not enough to fill the combs and then the empty parts of the wax cells become rotten, and since diseases gradually creep in, the honey becomes corrupted and the bees, too, themselves die. To prevent this, either the populations of two hives ought to be united, so that they can fill the waxen cells which are still sound, or, if a second swarm is not available, we must remove the honey-combs from the uninhabited parts, before they go rotten, with a very sharp knife. For it is very important also that a very blunt iron tool, because it does not easily penetrate, should not be pressed with great force and dislodge the honeycombs from their places; for if this has happened, the bees desert their abode.

There is also this cause of mortality among bees that sometimes very many flowers come up during several continuous years and the bees are more eager to make honey than to produce offspring. And so some people, whose knowledge of these matters is defective, are delighted at the large production of honey, not being aware of the destruction which is threatening the bees; for, exhausted by too much labor, very many of them are perishing and, as their numbers are not being increased by the addition of young stock, the rest at last die off. And so, if such a spring comes on that both the meadows and the cornfields abound in flowers, it is most expedient every third day to close the exits from the hives (small openings having been left through which the bees cannot pass), so that, called from the activity of making honey, since they have no hope of being able to fill up the waxen cells with liquid honey, they may fill them with offspring. Such then in general are the remedies for swarms suffering from some distemper.

XIV.

Next comes the management of bees throughout the year according to the excellent system set forth by the same Hyginus. From the first equinox, which takes place about the twenty-fourth of March in the eighth degree of the Ram, until the rising of the Pleiades, there are reckoned to be the forty-eight days of spring. During these days, he says, the bees ought to receive attention for the first time by opening the hives, so that all filth, which has collected during the winter season, may be removed, and, after the spiders, which rot the honey-combs, have been got rid of, the hives may be fumigated with smoke produced by burning ox-dung; for this smoke is particularly well suited to bees as if some affinity existed between it and them. The little worms also which are called moth-caterpillars and also the developed moths must be killed. These pests that generally adhere to the honeycombs fall off, if you mix ox's marrow with dung and, after setting the mixture on fire, bring the smell of burning near them. As a result of this precaution the swarms will be strengthened during the period which we have mentioned and will apply themselves to their work with more vigor.

But very great care must be taken by the man in charge, who feeds the bees, when he must handle the hives, that the day before he has abstained from sexual relations and does not approach them when drunk and only after washing himself, and that he abstain from all edibles which have a strong flavor, such as pickled fish and all the liquids which accompany them, and also from the acrimonious stench of garlic and onions and all other similar things. On the forty-eighth day after the vernal equinox, when the rising of the Pleiades takes place about the 8th of May, the swarms begin to increase in strength and number; but in the same period of days the swarms also which contain few and sickly bees die off, and at the same time in the extremities of the honey-combs bees are born of larger size than the rest, which some people think are king-bees. Some writers among the Greeks, however, call them oistroi from the fact that they excite the swarms and do not allow them any rest; therefore they recommend that they should be killed.

From the rising of the Pleiades to the solstice, which takes place at the end of June in about the eighth degree of the Crab, the hives generally swarm. This is a time at which they must be very strictly watched, so that the young brood may not escape. Then, when the solstice is passed and until the rising of the Dog-star, a period of about thirty days, the harvests of the cornfields and the honeycombs alike are gathered in. How the combs should be removed will be told presently when we give instructions for preparing honey.

Now Democritus, Mago and likewise Vergil have recorded that bees can be generated at this same time of year from a slain bullock. Mago indeed also asserts that the same thing may be done from the bellies of oxen, but I consider it superfluous to deal in more detail with this method, since I am in agreement with Celsus, who very wisely says that there is never such mortality among these creatures, that it is necessary to procure them by this means. But at this time and until the autumn equinox, the hives ought to be opened and fumigated every tenth day. This, though it annoys the swarm, is generally considered to be very wholesome. Then after they have been fumigated and are still heated the bees ought to be cooled by sprinkling the empty parts of the hives and pouring in water which is cold because it is very freshly drawn: then when there is anything which cannot be washed away, it must be cleansed with the feathers of an eagle or of any other large bird which are of a stiff quality. Moreover, caterpillars should be swept away and moths killed, which generally linger among the hives and are destructive to the bees; for they both gnaw at the waxen combs and from their dung breed worms which we call "hive-moths" Therefore, at the season when the mallows flower, when the moths are most numerous, if a bronze vessel of the shape of a milestone is placed amongst the hives in the evening and a light lowered to the bottom of it, the moths rush together from all sides and, flitting round the flame, are scorched because they cannot easily fly upwards from the narrow space or retire to a distance from the fire, since they are hemmed in by the brazen sides of the vessel. They are, therefore, consumed by the burning heat that is near them.

    About fifty days from the rising of the Dog-star is the rising of Arcturus, at which time the bees make their honey from the dew-drenched flowers of thyme and marjoram and savory. Honey of the finest quality is at its best at the autumn equinox, which falls before the first of October, when the sun reaches the eighth degree of Libra. But great care will have to be exercised between the rising of the Dog-star and that of Arcturus that the bees are not surprised by violent attacks from hornets, which generally lie in wait in front of the hives for them to come out. After the rising of Arcturus about the time of the equinox, which takes place when the sun is in the Balance (as I have said), the second extraction of honeycombs takes place. Then from the equinox, which occurs about September 24th, until the setting of the Pleiades, a period of forty days, the bees store up the honey that they have collected for winter food from the tamarisk flowers and woodland shrubs. Of this nothing at all must be extracted, lest the bees, disheartened by continual ill treatment and, as it were, in despair, should take to flight. From the setting of the Pleiades till the winter solstice, which falls about December 3rd in the eighth degree of Capricorn, the bees make use of the honey already stored up and are sustained by it until the rising of Arcturus. I am well acquainted with the reckoning of Hipparchus," which declares that the solstices and equinoxes occur not in the eighth but in the first degrees of the signs of the Zodiac; however, in these rural instructions I am now following the calendar of Eudoxus and Meto and the old astronomers, which are adapted to the public festivals, because this view, accepted in old times, is more familiar to farmers and, on the other hand, the subtlety of Hipparchus is not necessary for those of less refined education. On the first rising of the Pleiades it will be advisable immediately to open the hives and clear away any filth that there is and attend to them with particular care, since during the winter time it is not expedient to move or open the hives. For this reason, while there are some remains left of autumn, on a very sunny day, after the bees' habitations have been cleansed, the covers must be put inside close to the honey-combs to prevent there being any empty space within, so that the narrow quarters of the hives may warm up more easily during the winter. This must always be done also in those hives that are sparsely inhabited through lack of bee population.

Next any chinks or holes that there are we shall daub outside with a mixture of clay and ox-dung, and we shall only leave entrance by which they may come and go. Also, although a porch protects the hives, we shall nevertheless cover them by heaping stalks and leaves on the top of them and fortify them, as far as circumstances allow, against cold and bad weather. Some people kill birds and, after taking out their intestines, shut the birds up in the hives, so that in winter time they may provide a gentle heat for the bees which lurk amongst their feathers; furthermore, if the stock of food is used up, they can very well feed on these birds, if they are hungry, and leave nothing but the bones. But if the honeycombs supply their needs, the birds remain untouched, and do not offend the bees with their odor, fond though they are of cleanliness. It is better, however, in our opinion, when they are suffering from hunger in the winter time, to provide them with dried figs pounded and soaked in water or with boiled-down must or raisin-wine placed in little troughs at the very entrance to the hives; and it will be advisable to soak clean wool in these liquids, so that the bees, settling upon it, may draw up the juice as through a small pipe. We shall also do well to give them raisins sprinkled a little with water after we have broken them up. With these foods they must be sustained not only in winter but also at those seasons, when, as we said just now, spurge and also elms are in blossom. When the height of winter is passed, for a period of about forty days, they use up all the honey which is stored, unless an unusually generous allowance is left, and often too, after they have emptied the waxen cells, they lie fasting in the honey-combs in a torpid condition, like snakes, until about the rising of Arcturus, which is on the 13th of February, and by keeping quiet preserve the breath of life; in order, however, that they may not lose it, if too long a fast occurs, it is best to pour sweet liquids through the entrance of the porch by means of small pipes and thus support them during the temporary scarcity until the rising of Arcturus and the coming of the swallow with promise of more favorable weather for the future. And so, after this time, when the more cheerful weather allows it, the bees venture to go forth to their pastures; for after the spring equinox they are already roaming about everywhere without hesitation and plucking the produce of flowers which are suitable for the production of their young and carrying it into their dwellings.

These are the principles that Hyginus recommends for the most careful observation throughout the seasons of the year, but Celsus makes the following additions. He says that only in a few places are conditions so favorable as to provide different foods for the bees in winter and summer, and that, therefore, in places where suitable flowers are lacking after the season of spring, the swarms ought not to be left without being moved, but, when the spring foods are consumed, they should be transferred to places which can offer the bees a more liberal diet from the late-flowering blossoms of thyme, marjoram and savory. This, he says, is the practice both in the regions of Achaia, where the bees are transferred to pastures in Attica, and in Euboea, and also in the islands of the Cyclades, when they are transferred from other islands to Scyros, and likewise in Sicily, when they are moved from the other parts of the island to Hybla. The same writer says that the waxen cells are made from flowers and the honey from morning dew, and that, the pleasanter the material from which the wax is made, the better the quality which the honey acquires. He gives instructions to examine the hives carefully before transferring them and to remove honey-combs which are old and wormy and falling to pieces, and to keep only a few and these the best, so that as many as possible may be made at the same time from the better flowers. He also says that the hives that anyone wishes to transfer should only be moved at night and without being shaken.

XV.

Presently, when spring is over, as I have said, the harvesting of the honey follows. We conclude that the honey is ripe when we notice that the drones are being expelled and put to flight by the bees. They are insects of a larger growth, very like bees, but as Vergil says "a lazy herd" and idle, sitting near the honey-combs without doing any work; for they do not collect food but consume that which is brought in by others. Nevertheless these drones seem to contribute something to the procreation of the younger generation by sitting on the seeds from which the bees are formed, and so they are admitted on terms of some intimacy in order to sit upon the eggs which produce the new offspring; then, when the young bees are hatched, they are hustled out of the hives and, as the same poet says, "they are kept away from the fold." Some people recommend that they should be entirely exterminated; but I agree with Mago that this should not be done, but that a limit ought to be set to cruelty. For the race ought not to be wholly destroyed, lest the bees suffer from idleness, since, when the drones consume part of their provisions, they become more active in repairing their losses; but, on the other hand, a crowd of robbers ought not to be allowed to form a band, lest they plunder all the wealth of others. Therefore, when you see bees and drones frequently quarrelling with one another, you will open and inspect the hives, so that, if the honey-combs are half-full, they may be let alone for a time, but, if they are already full of liquid and sealed up with wax, just as if they had lids over them, the harvest of honey may be gathered in.

The morning should generally be chosen for the removal of the honey; for it is not advisable that the bees should be provoked when they are already exasperated by the midday heat. Two iron instruments are required for this operation, measuring a foot and a half or a little more, one of which should be an oblong knife with a broad edge on both sides and having a curved scraper at one extremity, and the other flat in front and very sharp, so that with the latter the honey-combs may be cut out better, and that with the former they be scraped off and any filth which has fallen upon them may be cleaned away. When the hive has been opened from the back, where there is no porch, we shall apply smoke made from galbanum or from dried dung; moreover, these ingredients are mixed with live coals and put into an earthenware vessel. This vessel has handles and is shaped like a narrow pot in such a way that one end of it is sharper through which the smoke may issue through a small aperture, while the other end is broader and has a rather wider mouth, so that the coals can be blown upon through it. When a pot of this kind is applied to a hive, the smoke is conveyed to the bees by the movement set up by the breath. The bees, unable to endure the smell of burning, immediately move to the front part of their abode and sometimes outside the porch. When there is an opportunity of inspecting the hives more freely, usually, if there are two swarms, two kinds of honey-combs are also found; for even if they live in harmony together, each community keeps to its own manner of shaping and constructing its waxen cells. All the combs, however, always hang down from the roofs of the hives, adhering very little to the sides and in such a way as not to touch the bottom, thus leaving a passage for the swarms. But the shape of the wax cells depends on the nature of the bee-house; for square and round and also long dimensions impose their own shapes upon the honeycombs as if they were molds, and that is why the honeycombs are not always found to be of the same shape. But of whatever kind they are, they should not all be removed; for at the first harvesting of honey, when the country still provides plenty of food, one-fifth of the honey-combs must be left; at the later harvesting, when the winter is already causing apprehension, a third part should be left. This, however, is not a fixed rule for all districts, since plans for the bees must be dependent on the abundance of flowers and the richness of the food available. If the hanging waxen cells run into length, the combs must be cut with the iron tool which resembles a knife and must be received by putting your two arms underneath them, and so removed; but if they run horizontally and keep close to the roofs of the hives, then you must use the scraping instrument, so that they may be cut down by the pressure exerted on the side which faces you. But old and defective honey-combs ought to be removed, and those which are soundest and full of honey should be left, as also those which contain young bees, so that they may be preserved for propagating a swarm.

Next the whole store of honey-combs must be collected in the place where you intend to make the honey, and the holes in the walls and windows must be carefully daubed over, so that there may be no passage for the bees which obstinately search as if they were looking for lost wealth, and, if they track down the honey, eat it up. Smoke must, therefore, also be kindled of the same materials as before at the entrance of the place to drive away those that are trying to get in. Then those hives from which the honey has been cut out, if they have combs lying across the entrance, will have to be turned round, so that the hinder parts in their turn become entrances; for in this way, the next time the honey is taken, the old combs rather than the new will be removed, and the waxen cells, which deteriorate as they grow older, will be renewed. But if the hives happen to be surrounded by walls and cannot be moved, we must take care that the combs are cut out, sometimes from the back and sometimes from the front. This process will have to be carried out before the fifth hour of the day and then repeated after the ninth hour or else next morning. But whatever be the number of honeycombs that are harvested, you should make the honey on the same day, while they are still warm. A wickerwork basket or a bag rather loosely woven of fine withies in the shape of an inverted cone, like that through which wine is strained, is hung up in a dark place, and then the honeycombs are heaped in it one by one. But care must be taken that those parts of the waxen cells, which contain either young bees or dirty red matter are separated from them, for they have an ill flavor and corrupt the honey with their juice. Then, when the honey has been strained and has flowed down into the basin put underneath to catch it, it is transferred to earthenware vessels which are left open for a few days until the fresh produce ceases to ferment; and it must be frequently skimmed with a ladle. Next the fragments of the honeycombs, which have remained in the bag, are handled again and the juice squeezed out of them. What flows from them is honey of the second quality and is stored apart by the more careful people, lest any of the honey of the best flavor deteriorate by having this brought into contact with it.

XVI.

The yield of wax, though of little monetary value, must not be overlooked, since its use is necessary for many purposes.

The remains of the honeycombs, when they have been well squeezed, after being carefully washed in fresh water, are thrown into a brazen vessel; water is then added to them and they are melted over a fire. When this has been done, the wax is poured out and strained through straw or rushes. It is then boiled over again a second time in the same manner and poured in such molds as one has thought suitable, water having been first added. When the wax has hardened, it is easy to take it out, since the liquid that remains in the bottom does not allow it to stick to the molds.

www.ingramcontent.com/pod-product-compliance
Lightning Source LLC
Chambersburg PA
CBHW031553210526
45464CB00003B/1281